2018 e-book release
ISBN: 978-0-6482298-4-1

2018 paperback release
ISBN: 978-0-6482298-5-8

Author: Ken Walker
All photos supplied by Ken Walker from his personal collection.

Registration:
Thorpe-Bowker
Level 1, 607 St Kilda Road
Melbourne, VIC 3000, Australia
tel: +61 3 8517 8342
e-mail: bowkerlink@thorpe.com.au

First released and printed in 2018

KEN WALKER

FROM AIR CADET TO LANCASTER BOMBER PILOT

"The Autobiography"

CONTENTS

Copyright		2
Prologue		6
Chapter 1:	*Prelude To War*	8
Chapter 2:	*The Air Training Corp - WWII*	12
Chapter 3:	*The Aerodrome Defence Unit*	16
Chapter 4:	*Secondary Flying Training (Canada)*	19
Chapter 5:	*Operational Training Unit (England)*	28
Chapter 6:	*The Real Business*	36
Chapter 7:	*On 'Ops'*	45
Highlights In My Journeying		74
The Aftermath		77
Some Terms In Common Air Force Jargon During WWII		79
Pilot's Flying Log-book		81
Ken Walker's WWII Medals		98

PROLOGUE

Kenneth Hollway Walker arrived on this planet on 4th February 1924. The place of that arrival was the city of Dunedin in the south Island of New Zealand.

This is a brief introduction into which, hopefully, will be the story of the events he experienced during World War II, along with thousands of others.

Many friends have suggested that he make this record, as he not only lived through, but also participated, the making of what was a major sequence of world history.

But first, in 1929, the great world-wide depression commenced, continuing until well into 1935, on the 28th of November of which his father, at the age of 51, died, which he believes was the direct result of that completely unnecessary man-made monetary scourge. However, that is another story, not to be aligned to this story herein, but should be in another recorded narrative.

Ken's primary education was at High Street School in Dunedin, his secondary was at Otago Boys' High School, 1937 to 1940 inclusively, at the end of which he sat for, and passed, his University Entrance examination. His widowed mother, on her aged pension, could not have afforded to have him go to university.

Thus, at the beginning of 1941 he obtained his first employment in a well-established company learning clerical and record keeping work, which stood him in good stead in much later post war years. Now let's get on with the motley shall we…

CHAPTER 1: PRELUDE TO WAR

At the start of 1941, the war was going dreadfully for Britain. Hitler's Nazi Germany had overrun and occupied most of the European countries. Our ally, (?) France had been beaten, along with British army and Royal Air Force units which had been dispatched to France. Britain was, with the assistance of early units from Empire countries, alone facing expected invasion by Germany.

During the pre-war years, despite the obvious German rearmament, the pathetic attitude of British governments under first, Stanley Baldwin and then the nincompoop Neville Chamberlain had ignored the plainly obvious threat Hitler was posing, and failed to also modernise and increase its own defensive armed forces.

Fortunately, even though he had been warning his own party in government that Hitler was power hungry, thus posing an obvious threat to all European countries as well as Great Britain, Winston Churchill was recalled from his parliamentary back bench position to become the party leader and the Prime minister. Hence Britain was brought aware immediately, war production was not just increased from its World War I form, now outdated, in particular for the air war, for which Nazi Germany had equipped itself and used by bombing first Poland, then other European countries.

It was only good fortune that production of the two new fighter aircraft had been stepped up and put into squadron service, the Hurricane and the Spitfire (affectionately referred to as the 'Spit'). Each of those two aircraft was powered by Rolls Royce Merlin engines.

Hitler boasted that he was going to destroy before invading and occupying Britain. Thus began the 'Blitzkrieg' (which was combining two German words, blitz meaning lightning, and krieg, meaning war- so it was to be a lightning war).

Massive armadas of German two engine bombers began attacking Britain from early 1941. However, the Spits and Hurricanes became airborne daily, successfully defeating the blitzkrieg bombing formations.

Then, one day in September 1941, the fighter pilots, as was the custom, were standing by in the ready for the call to scramble, the code word to get into their aircraft and take off to meet the enemy. But on that day the call to scramble was not made. The German bombers were being shot down in such numbers that Hitler called off the day bombing. Thereby he suffered his first ever defeat.

That "Battle of Britain" has become a celebrated historical event. This episode, I have deemed, in my opinion necessary to repeat, as it is order of sequence to the essence of the following history, which is relevant to the role to come of Bomber Command. It is also important to remember now that hundreds of men from the Empire countries had gone over to Britain, serving and flying in those early war days. Many thousands followed to be engaged in the strategic bomber offensive yet to come.

At war's beginning, the officer in command of Bomber command was Air Chief Marshal Sir Cyril Newall. He was a square peg in a round hole. He had no conception of how the bomber force could, and should have been developed. Churchill rapidly moved him sideways to be the Governor General of New Zealand, where he quickly made himself unpopular.

Churchill's replacement to Officer in Command was Air Chief Marshal Arthur Harris, who had served in France in World War I. He remained in the RAF Post war. He rapidly became a very effective commander, becoming in charge of RAF units in India and adjoining regions. He had worked out that if war was again to break out, night bombing would be the most successful to defeat any prospective enemy.

In those years his brain was in overdrive. His thoughts and decisions were about everything required to build up an effective plan, whereby night bombing could be successful. Among the vital needs were new more modern aircraft to be specially designed and equipped with the correct tools and instrumentation, plus intensive pilot training in night air navigation. Sadly, governments being peace minded did not provide the aircraft types he needed, but he did proceed with what aircraft he had to introduce the training as herein. That proved to be a wonderful success, on which he was to base his new World War II command.

CHAPTER 2: THE AIR TRAINING CORP - WWII

Churchill placed full reliance in Harris who outlined his stated requirement as above, particularly larger heavy bombers, which could take much larger bomb loads, because at that stage, a strategic plan to attack Germany's war production targets was the only avenue open to take the war to Germany. The only heavy bomber now in service was the Vickers Wellington which was not able to take heavy enough bomb loads.

Absolutely the first priority was setting up an Empire wide plan for the intensive training of personnel to crew the huge bombing groups to come. Thus, the 'Empire Air Training Scheme' (EATS) was put in place. This was promptly inaugurated mainly in Canada, Australia, New Zealand and Rhodesia, all of which were remote from the theatre, hence no interruptions to the training schedules. Canada, because of its vast land area, provided ideal training, not only for its own airmen, but also for many of the recruits from the other countries, including from the UK.

In addition, when Harris took over he inaugurated the 'Air Training Corp' (ATC). This enabled young men with good academic standards who had not yet attained the age of 18 years, which was the minimum age to be eligible to be enlisted in to all the armed forces, to join the ATC. The ATC started in Britain in February 1941 and in New Zealand in May 1941. This was because there were many 16 and 17 year old lads who were able to join this cadet movement to learn much of the aircrew subjects at that early time before getting into the Air Force at 18 years.

In New Zealand at aged 17, I enlisted into the ATC in May 1941, and being accepted, we were all issued with our Air Force uniforms. The jackets, instead of having brass buttons, had black Bakelite buttons. Two nights each week we reported to a commercial building generously granted by the business owners. Instructors from the nearby RNZAF station came to teach us the ground subjects needed, such as air navigation, theory of flight, Morse code, signalling and ancillary parts of those.

We also had a Link trainer, the pioneer of what has developed as the modern flight simulators. In the Link we were taught how to operate the pilot's job, needed to fly an aircraft, by learning how the controls worked the connection to the gearing of the control surface on the aircraft. Also we underwent drill which is generally referred to as 'square bashing'. I had already done that in my four years at secondary school where, every Monday morning, we wore army cadet gear and we

Self in uniform of ATC Air Cadet with my brother Jack who was in the NZ Army. Photo taken in 1941 at front door of our home, 127 Stafford Street Dunedin.

were drilled.

I must have shown good results, being promoted to the rank of Corporal. Two months later I became a sergeant. Our group was split into two ‘flights’, each the equivalent to an army platoon, the sergeant of each given the task of drilling his flight. In all services this is learning discipline which is always compulsory which I found out as time went by.

Sometimes we went out on public streets combining some manoeuvre with marching. Occasionally people who were observing us would applaud, as well as getting to know who and what we were. Newspapers recorded our existence. We did learn a lot which was a fine springboard for what the future held for us. I thoroughly enjoyed that first stage.

CHAPTER 3: THE AERODROME DEFENCE UNIT

The first call up to the RNZAF was that those of us ATC cadets now aged 18 years were enlisted into an Aerodrome Defence Unit, a new unit which in effect was a holding phase, until actual aircrew training schedules were ready to receive us. At the ADU, the bakelite buttons on our cadet tunics were removed with brass buttons replacing them.

I was posted to the air force station at Taieri (pronounced Tie-ree), which was a short distance south of Dunedin. We were then issued with 030 gauge rifles, a Yankee weapon, with which we were to defend the station in the event that some enemy force had invaded New Zealand. (Please stop laughing). At that time, no such enemy existed, thankfully. We went through daily drilling, also classes to inform us about all Air Force protocols. That phase lasted about two months.

Our first posting to our aircrew training was to an 'Initial Training Wing' (ITW), which was held at the North Island city of Rotorua which is the centre of the geothermal area. Pre-war, it was the most popular of the NZ tourist attractions. During the war the government took over the entire city for the Air Force to be able train aircrew students. We were billeted in the hotels and I was in

Brent's, a former top hotel. I was in a two man room along with a Scottish man "Scotty" Walker, much older than I. He was like a father and mother to me. Previously, I had never been out of the South Island.

The course at ITW was ground subjects only, some of which we had covered in Air Cadets, now with the addition of armaments – accent on machine guns as fitted to aircraft turrets in bombers, Also we were tutored in the internal combustion engine. Physical and medical examinations were often and thoroughly carried out, predominantly to determine which position in a crew each student would eventually be suited for.

For a pilot's role special tests were devised for eyes, to ascertain depth perception, to find out if you were landing an aircraft you would be liable to set it down on land when you were still too high off the ground. Also what were then known as field tests, in order to determine if you became aware of anything else all around you at different distances, such as enemy aircraft. Hence, these two special eye tests would prove if you were qualified for the pilot's position.

I did pass this course, both academically and health wise, being posted back down to Taieri where I had done my ADU stint, thus I then started the first actual flying course. This was in the old twin wing Tiger Moth. This was called Elementary Flying Training School (EFTS).

The average number of hours flying with the instructor before he deemed his student was ready to fly his first solo circuit was between eight and nine hours. I took eight and a half hours, when my instructor, Pilot Officer Payton, sent me off by myself. No one could ever forget their first "solo". It was a mixture of nerves and exhilaration. When I landed, which I was told was a perfect one, and that I would probably never again repeat, none the less, I felt quite a lot of pride.

Flying in the open cockpit of the "Tiger" with the wind in one's face was, to me, very enjoyable. It was not too long till I 'got the hang of it'. For me it was both exciting and satisfying. This was what I wanted to do. It was a thrill at the end of the course when I was informed that I had easily passed.

CHAPTER 4: SECONDARY FLYING TRAINING (CANADA)

On completion of the EFTS, half of those who had passed were posted to the next stage of flying training, "Secondary Flying Training Course" on a NZ station, the other half, of whom I was one, were posted to do SFTS in Canada.

Thus in the early months of 1943, our group boarded a troop ship, the Matsonia, an American pre-war cruise ship, now commandeered to be a troopship. America was now in the war after the Japanese had "blown their bugle' for them at Pearl Harbour in December 1941.

For me, it was a thrill being at sea. Also on board were many American servicemen who had been wounded in action against the Japanese. Our first port of call was at San Diego where all the Americans were taken ashore. We Kiwis were forbidden to go ashore, even for an hour or two, because, we were informed, San Diego was notorious for street fighting between American army personnel and their own navy sailors.

The following day, the ship left harbour sailing north to San Francisco, where it berthed. We were notified that in about six hours we were to be taken north to Vancouver, the western port of Canada, by an American train. We were somewhat taken aback when we were seated for meals in the dining carriage. Waiters, mostly Afro-Americans, brought our meals and they placed an empty bowl on the table, which was placed there with a flourish. We learned this was a Yankee custom for diners to put their 'tips' in. Evidently their wages were low, they relied on tips, poor fellows. Our pay was still quite meagre, so we were not tipping. We were not exactly at the top of the popularity poll.

Now a highlight. We boarded a Canadian Pacific line train. That journey over the Canadian Rocky Mountains was so remarkable, not just for the majestic scenery, but also for the unbelievable engineering involved in its construction, which was in the times before the age of heavy earthmoving and other modern equipment came into existence. When the train was at Banff, by the beautiful Lake Louise, for a brief stop only, we had no chance to get off even for five minutes. More flying training beckoned.

We were posted to Royal Canadian Air Force (RCAF) station at Yorkton in the province Saskatchewan, the third most western province in Canada from Vancouver. On arrival at Yorkton we were a week early before the present course graduated, and the next one (ours), was due to begin. On that due day we were present for the roll call of students making up this new course. But not one of our

Kiwi group had his name called. Panic stations!

Urgent signals to RNZAF headquarters Canada in Ottawa, where it was discovered that an internal staff mistake had been made. We were intended to be at RCAF Claresholm Alberta province. Bureaucracy is not a modern development; it was already flourishing in 1943!

Two days later we arrived at Claresholm, two days late for the start of our course, for which we were extremely unpopular.

Never mind, we did start. This SFTS course training aircraft was the twin engine Avro Anson, an English kite fitted with American made Jacobs engines. One of the six basic flying instruments – the turn and bank indicator – was also a Yankee one different from the British design we had on the Tiger, however, we became used to it.

Each instructor had two students allocated to him. My student partner was Bert Jeeves from Lower Hutt and our instructor was Sergeant Wilf Crough, a Canadian. Wilf was fresh out of Instructors' course; hence Bert and I were his very first students for him to train. I sensed that Wilf was slightly tentative when he was explaining the characteristics of having two engines instead of one. Wilf did settle down and I found his instructing became very plain to understand, also very easy to carry out all he was teaching.

It is, as in all things, very important to pay attention closely to what is being taught. I was always conscious of the four Latin words, the motto of Otago Boys' High School – 'Recti Cultus Pectora Roborant', in simple English "correct learning strengthens the mind". I continued to apply that motto in all subsequent flying.

Bert also told me that he had similar thoughts about Wilf's early days in his training, and as well how he did settle down. We both passed our final academic and flying tests.

We were both on the graduation parade on 15 September 1943 at which we were promoted to Sergeant. I still have the photo of the Station Commander, Group Captain Kennedy, attaching my pilot's wings on my uniform jacket. An auspicious occasion! The Group Captain was a very fine officer and a true gentleman. We found Canadians to be most generous in their hospitality to all of us from the other Empire countries.

It is important to mention that there were many UK students on the course, because in the UK land for early training was very limited.

Station Commander, Group Captain Kennedy, attaching my pilot's wings on my uniform jacket. 15 September 1943.

Next stage, over to the east coast port of Halifax, the embarkation station where we stayed until our time to board ship to go to the UK. During our stay at that RCAF station we underwent rigorous medical exams and other tests such as night time aircraft recognition of our own and German planes.

Late one afternoon our group was rapidly assembled and we were notified we were to be taken to the port to board ship. We expected that this would be to the east, to the marine docks of Halifax to get on a troop ship. However, when the train on which we were bundled started travelling west, none of the officers in charge of us told us anything. Were we on a mystery journey? War time security was very strict.

We finally arrived at Montreal and quickly transferred to another train which was pointing to the south. This had us confused. There were not any civilians in the carriages we were seated in. We took off and it was not until we were in the USA that the officers in charge of us came through the carriages to inform us we were going to the port of New York, where we would board 'a' troop ship to take us to the UK, and that we would not be sailing in convoy.

Arrived at New York, boarded the ship – would you believe, dear reader, it was the Queen Mary, the first of the large pre-war passenger and tourist ships. It, like some others, had been commandeered for use in the war effort.

As the ship was eased out of where it had been berthed, and was sailing toward the harbour exit, surprise, surprise! At that moment the Queen Elizabeth had entered the harbour. It was the very first time that these two grand ships had been in that harbour at the same time. The New York large fire-fighting boats sailed out, greeting these two great vessels with a fine display with their huge deluge nozzles sending a wonderful stream of water as a friendly greeting to them.

The reason these large ships did not sail in convoys was their own great speed. All convoys had to sail at the speed of the slowest ship for obvious reasons. The German U boats (submarines) were sinking so many merchant ships, despite naval escorts. For a U boat commander sighting a ship, whether or not one in a convoy, he needed to use the U boat's instruments to calculate that target's course and speed when looking through the periscope, while knowing his own boat's speed and course before he gave the order to fire a torpedo. This took an interval of time, which at my age now, I am unable to recall, but it was from the moment of his first sighting of the target ship to the time he gave the order to fire. This interval was recognised in Britain, thus the speed of the Queen Mary was much higher than that of the submerged U boat, thus the Queen Mary and the Queen Elizabeth zigzagged their course before the time interval, as explained above, expired. This constant course changing defeated the U boat commander's calculations.

For us, it was thrilling with this never ceasing zigzagging. One moment, when out on deck with hands held firmly held on a railing, we would be looking out to sea. Then, in a short time, the Mary changed course, and that sea would be on a higher level above us. Schoolboy stories seemed to be coming true. Exciting stuff!

About six or seven days later (from fading memory) the Mary arrived off the south west tip of England. Two Mark 5 (clipped win) Spitfires, flying in a tight formation, came out to greet and welcome us - my very first sighting of this famous fighter. This flying display of that close formation continued for about five minutes. It had to be seen to be believed.

However, all good things come to an end. The Mary changes course, sailing north through the Irish Sea – no shamrocks in sight. Late that night the ship berthed at the Scottish city of Greenoch on that country's west coast.

We were rapidly disembarked, and escorted on to a waiting train, which then took us down to Brighton, a famous pre-war holiday centre. This was the receiving centre for both Aussie and New Zealand newly graduated aircrew. The Aussies were billeted in the Metropole Hotel, we Kiwis on the Grand Hotel. Our time there was for when the next, and more intense, training was to be ready to accept us.

Chapter 5: Operational Training Unit (England)

In the meantime, I was posted to RAF Wheaton Aston in Staffordshire, where we were flying in the light twin Airspeed Oxford. On that brief course we were tutored on a blind flying landing procedure based around two beacons giving out signals. It was interesting but in times yet to come, I was not ever called upon to need to use it. From there I was posted to Silverstone RAF station in Northamptonshire which these days, is the location of the English leg of the world formula one motor racing championship.

We are now starting in the real serious series of our bombing training. This was 'Operational Training Unit (OTU). It was here that all aircrew "trades" came together. All of us the day after we arrived there, were brought in to a hall, where we were addressed by the Station Adjutant, a non-flying pukka Englishman, and administrative wallah, who briefed us on most of the nearby places, one of which was a village by the name of, at which he took a piece of chalk and wrote on a blackboard: TOWCESTER which he then announced: "for the benefit of our colonial personnel is pronounced "TOASTER".

He then, quite innocently asked us all, "Now how is it pronounced?" in reply to which a loud chorus of Australian accents blurted out, "TOW (as like cow) cester sir". This raised the hackles of the 'Adj' and the following five or six minutes was taken up with an argumentative cross yelling about this orchestrated Aussie style of 'taking the mickey' with the poor - now very heated – man.

However, after that 'concert hall performance' came to a conclusion, he told us all to go outside where the pilots would select their own crews from the other 'tradesmen'. I was standing on the right hand side of a car beside its bonnet when I became aware of the fingers of a man standing on the left side of the car strumming on the bonnet, noticing that his fingers were very long and supple. Also I saw his half wing flying wing had the initial S; a wireless operator. I asked him, "Would you be on my crew?" His reply: "Some poor bugger will have to put up with me so it may as well be you". Norman became a fantastic wireless operator.

As I then moved away, Norman in tow, I saw a nice looking bloke who was an air gunner. On the spur of the moment I asked him, "Would you like to be one of my gunners?' to which he agreed saying "I think it would be good to have a Kiwi skipper". He then told me he had a good mate with whom he had been through all their training. Together we found his mate, thus Joe and Alf became my gunners; Joe rear gunner, Alf the mid upper one. That left two more to find. I strolled on

My crew - A 44 (Rhodesia) Squadron, Spilsby Lincolnshire, England
No 5 Group Bomber Command

Back row left to right: Jack Hardy (bomb aimer) Jock Spence (mid-upper gunner) Cyril Crooks (Flight engineer) Joe Ward (rear gunner)
Front row left to right: Norman Roberts (wireless operator) Self, Alan Turner (Canada, navigator)

around, sighting a fellow with an N for navigator, also the word Canada on his shoulder. He accepted my invitation to join my crew. Now only a bomb aimer to find. Derrick came on to the crew and it appeared we were among the first few crews to form. That evening all six of us were allocated beds in the same Nissen hut, our sleeping quarters.

Next morning we met the pilot who was to be our flying instructor. He wanted to get to know each one of us personally. That impressed all of us. He, like most of the instructors had, with his crew, already completed an actual 'tour' of bombing operations (ops), then, after a well-earned leave was posted to a flying instructor course. Now, having passed that, he was an instructor. We were not his first crew to train. Each crew member had also to get extra 'operational flying training' with other experienced tutors, in other aircraft, which was deemed essential for our flying training as a crew.

This time, flying as a crew was of prime importance. Our aircraft was the Vickers Wellington, affectionately referred to as the 'Wimpy' which I seem to recall was the name given to a man named 'Wimpey Wellington' in a newspaper cartoon. The Wimpey's fuselage ('body' to the 'uninitiated') was of what was named to be of a 'geodetic' shape, designed by a well-known aircraft designer, Barnes Wallace (more about him later). It was quite a nice craft to fly. It has also been the first of the twin engine 'heavy' bombers, now used in training and some other commands.

Accent on this course was training for night flying which meant intense flying on instruments because the strategic bombing offensive, as designed and directed by Air Chief Marshal Harris was based on night attacks continually on German war production and supply in many forms. This was the only practical manner available to our British forces in which the war could be taken to Germany.

Hence I learned to concentrate my learning on each and every aspect of instrument piloting. The absolute importance and success depended on this night flying ability. This applied equally to each and every crew member. That entailed every crew becoming a united and harmonious unit, a band of brothers, not just while flying, but also in our social life. Yes, we did ultimately have some of that occasionally. We all depended on each other.

I mention here that we did not have just one instructor, but quite a number. The individual crew members each had both ground and flying instruction on their 'trade' responsibilities.

The gunners had flying time in which their tuition was, when flying, practising shooting at moving targets (such as enemy aircraft). Those targets were called drouges, long hollow material attached to a metal ring towed behind an aircraft to test the gunners' deflection shooting skills. The other crew blokes, e.g. the navigator, had to do the navigation on short and long trips. The bomb aimer dropped 'target bombs' on a marked target on a remote field. The wireless operator received radio instructions which he had to pass on to the pilot or, on some occasions, to the navigator. Norman was top grade in his craft.

This OTU was a multi-tasking level of instruction. We, as a crew, passed the course, which was lasted one and a half months. It was highly concentrated both mentally and physically. During that time we, as a crew carried out what was referred to as a 'bullseye operation', which we called 'spoof ops'. This was to fly a diversionary flight. On each occasion, both at night, when an actually real bombing raid by 'operational' squadrons, we, not with an instructor on board, were briefed to fly with other training command aircraft to proceed over the north sea to the west of the German coast in order that the German radar could detect us, in such numbers, to convince them that we formed an actual bombing stream on the way to a target.

RAF designed this to get German night fighters airborne and vectored towards us. Then, when timing planned by Bomber Command headquarters we were instructed by wireless code to turn around, drop height and return to base. The overall purpose was to get the Jerry fighters flying away from the real bombers and out of the area of the true target. This caused the German fighters to use much of their fuel, thus by the time we had fled the area they were in need of refuelling, before they could 'scramble' again to get to the actual bombers which by then were well on their way back home. Wartime security was, as previously noted, so strict that we were prevented from discussing anything at any place. However we did get to hear that our 'spoof' ops did achieve success. We also got quite a buzz at having gone on a 'near op'.

Now, we were posted to RAF Wigsley in Lincolnshire for 1654HCU (Heavy Conversion Unit), where we were to convert from the twin engined aircraft to four engined ones. Here we also got also our seventh and last crew member, the flight engineer. They were not previously needed on the twin engine aircraft, but now with four engines they were truly needed.

The aircraft we trained on now was the Short Stirling which was the first 'heavy' to go into operational service in the RAF. It had, by now, been taken out of 'ops' being replaced by the Handley Page Halifax and the Avro Lancaster. However the Stirling was still playing a valuable role. There were seven fuel tanks in each wing which required the engineer to manage the correct tank or tanks selection. His name was Cyril Crooks, then aged nineteen. He became most capable in his trade.

Again, we had to concentrate on instrument flying and procedure when an engine failed because in operations, fighter attack often resulted in an engine being shot out. So much more for a pilot to learn with a four engined 'kite'. More emergency actions were a high priority for our instructors to teach us.

That course took one month. I saw in my logbook at the end of that course, in the summary of our crew I was signed off as a pilot suitable to be a heavy bomber pilot. The signature of the Chief Flying Officer was R. Kingsford Smith. I seem to remember that he was a cousin of the world famous Australian Sir Charles Kingsford Smith.

Next stop was proceeding to RAF Syerston in the county of Nottinghamshire (no sign of Robin Hood or the Sheriff). This was Lancaster finishing school, converting to the famous of all bomber aircraft in World War II. We spent six days 'converting' to the craft in which we would soon be flying in an 'operational squadron'. The Lanc (its title of affection) was powered by four of the great Rolls Royce Merlin engines. At that point I was promoted to the commissioned officer rank of Flying Officer.

CHAPTER 6: THE REAL BUSINESS

We were posted to RAF station Spilsby in Lincolnshire where we became a 44 (Rhodesia) squadron crew. As in all operational squadrons, many of our crews were a mixture of men from different Empire countries because of the Empire Air Training scheme. Despite the name Rhodesia, our crew was one such.

However, the day after we had arrived at Spilsby, our mid-upper gunner, Alf Wisson, became sick, reporting to the medical officer (MO – station doctor), who, after examining him completely diagnosed sinusitis, a condition in which he was no longer allowed to fly. This was because sinusitis does not allow air to pass between the inner and outer ear and as we climbed to higher altitudes the air pressure lessened, and increased again when descending This would cause bursting of the ear drum, thus poor Alf was 'scrubbed' from flying. He left Spilsby immediately, we knew not to where, but probably as an instructor (non-flying) at an air gunner instruction school.

That left us one man short. However, luck came to us. Some six months earlier, one 44 Squadron Lancaster had been attacked on a bombing operation by a German night fighter, suffering quite a lot of damage, with their mid-upper gunner being badly wounded. Fortunately the pilot succeeded in nursing his damaged Lanc back to a good landing. The wounded gunner was carefully removed by the waiting ambulance crew. He was quickly transported to a hospital where he received great medical care for six months, at the end of which he received doctors' clearance, certifying he was now fully capable to resume operational flying.

In the interim period, his first crew, with another gunner, had continued to keep flying, completely finishing their 'tour' of thirty ops, going on leave, and probably posted to training command as instructors. Therefore Bill instantly became part of our crew, replacing Alf Wisson. Bill's already experience of ten ops was valuable to us. He being a Scotsman was, of course, 'Jock' to us. Also, he was one more 'other country' bod, (the Air Force abbreviation for body — or person). On reflection, that was a precursor for this modern day, in my opinion, bloody superfluous political correctness. Sorry dear reader for that outburst. I had better resume with the object of this discourse.

The initiation for a newly arrived crew to an operational squadron was a flight teat with the squadron instructor. This man was Flight Lieutenant Webster (pronounced 'Leftenant', not like the Yankee bastardisation 'Lootenant', thinking it sounds as though it means the occupant of a lavatory).

'Mr' Webster was a Rhodesian who was a stern, humourless fellow, quite the opposite as we were soon to discover. Most Rhodesians were very fine men, very likeable blokes (I still like that word) with a good sense of humour. Back to our test flight: 'Mr' Webster gave us half an hour in which he gave me orders to carry out a variety of turns and manoeuvres, some in a quick moment. I presumed he was testing my reactions to emergency situations, all the while observing the flight engineer's performances. He eventually told me to land.

I knew he would have to assess my landing procedure from my approach to my actual touch down on the runway. I recall my confidence at that time, "I am going to make a perfect landing". So I did. That over, 'Mr' Webster pronounced, reluctantly, I thought, that we were fit to fly on 'ops'. First hurdle successfully cleared. Next step: every new crew pilot was to fly with an experienced crew on an actual bombing raid. This was referred to as doing your 'second dicky'.

For this I flew with an Englishman, Flight Lieutenant Davey and his crew. Before we were even driven to his Lanc, L-Love, he spent time with me, explaining what I was about to observe, also for me to pay close attention to the verbal discipline to be observed by all crew when talking on the crew intercom and with each other. Each man must name his crew position when going on the intercom, no personal names used and no unnecessary chatter. He impressed on me the vitally important moment when we commenced the 'bombing run' when only the bomb aimer and the pilot were talking. The bomb aimer was lying prone in his position in the nose, eyes on his aiming sight, tracking our way to the target marker which had been dropped by Pathfinder Force crews.

As we approached bomb release time, if the aircraft went off its track to the target, the bomb aimer would tell he pilot "left", or "right", then when back on track he would say "steady, steady" until the cross hairs on his bombsight were on the target marker. He then pressed the button on the end of the bomb release tube.

F/L Davey's explanation prior to taking off had been so well explained, I thanked him. His response was "I was a new bod myself once, and mostly I found Kiwis usually became good pilots" That boosted my confidence.

On route to the target, we first encountered ack ack (anti-aircraft shooting) from German flack ships off the coast of Europe, then, on crossing the coast, land based ack ack concentrations also opened fire at us. F/L Davey seemed to thread his way between them. I was standing just to his right side with my helmet intercom lead plugged into a nearby jack.

It quickly became obvious to me the reason we had to have such intense training. We are now in this new world of all-out war. But now, we were taking the war to the Nazis. Originally, they were bringing the war to us.

As this flight continued, German night fighters were spotted getting into our 'bomber stream'. I saw two of our aircraft on fire - not a healthy sight. But our mob 'pressed on regardless'.

Finally we reached our target, the Mittelland canal at Gravenhorst. Germany was producing submarines (U boats) in sections at inland factories, then placing them on barges which were shipped on these canals down to be connected to each other at highly defended ports. Hence, canals were attacked to try and cause sufficient damage by taking them out of service. F/L Davey, at suitable intervals, gave me a running commentary over the intercom, with explanations and advice that I programmed into my computer (the brain is a computer). It was that high school motto at work again.

Our flight returning to base was interrupted when the wireless operator received a signal that, because thick fog had descended over our Spilsby base, we were diverted to land at a drome named Luffenham. The skipper asked the navigator to look in his manual to find out what the base was used as and to work out a new course to get there. We were still over the North Sea. The navigator told Skip that Luffenham was a lot south of Spilsby and it was an American station, giving a course to it and adding cheerily, "American base, good-oh, ice cream, chocolates, peanuts!". English food rationing was so tight, but the Yanks always had these luxuries in plentiful supply. But these things about to be anticipated were torn from our thoughts because suddenly our kite received two quick hits from some weapon which we could not understand.

The skipper ordered the flight engineer to ascertain what damage had been sustained. He found one of the hollow aluminium rods that went down the port (left) side to a control surface, a rudder, had been hit, and that for about twelve inches of its length had been shot away. All but about half of its circumference was left holding it together. There was nothing on board which could make even a temporary repair. There were, from my memory, two or three shrapnel holes in the fuselage which could affect flying ability.

The skipper, taking account of the damage to the control surface being quite dicey, decided to head for and land at one of the three emergency landing fields up the eastern side of England. They were at Manston, Woodridge and Carnaby. Each one had a longer runway than on any of the operational stations, with the added safety of a 500 yards grass undershoot and overshoot areas. Skip decided to go to Woodridge. He took great care to fly gently in order to not place too much strain on the damaged control rod. His approach and landing was masterful.

He was then guided by ground staff, taxying to a parking area. Then he reported to control tower, giving details of our squadron and his name, asking for a signal to be sent to Spilsby notifying our arrival at Woodridge. Skip looked completely drained by now. However, we were all taken to a mess (dining hall), being given a fine 'operational meal' after which we were taken sleeping quarters. After that night of variety sleep came easily.

The next morning a complete examination of the damage commenced, and we were then free to do as we pleased. A Yankee B24 Liberator 4 engined bomber had also landed there the day before. F/L Davey and I came across the second pilot of that B24'Lib' and we got talking. The Yank invited us to come and take a look over his bomber. The two American 4 engined bombers, this Liberator and the B17, named the Flying Fortress, each had a crew of ten men, which included two pilots as against our British 'heavies' seven men crew, with only one pilot. The controls and ancillary equipment of the B24 was a mass of electric switches for a range of controls. Our British planes' controls included hydraulic and manual levers.

The Yank was typically most enthusiastic about his 'Lib'. He accepted F/L Davey's invitation to come over and enter our Lanc. He was astounded at the lack of space in the Lanc, especially on the flight deck. He found it unbelievable that we had only one pilot. In his southern Yankee drawl I remember he remarked, "I can't make out how this goddamned thing can get up off the ground". Neither of us commented on his words.

Later that day, one of our '44' pilots with only his flight engineer, F/O Daggart, arrived at Woodridge, to ferry us back to Spilsby. After all damage to F/L Davey's Lanc had been fully repaired and flight tested, being passed as fit again for operational flying, it was flown back to Spilsby.

It appeared that ammunition which had hit our aircraft, as mentioned previously, was due to the time interval between our changing course to fly to Luffenham, and our being fired at. RAF headquarters needed to signal Royal Navy HQ that our bombers would now be crossing the coast at a point a lot to the south of that which Navy had been briefed we would be. Hence, alert gunners on one of a naval unit escorting a cargo ship were unable to visually identify us through the cloud cover. Thus we were hit by friendly fire. What an initiation to my 'operational piloting'.

My crew bods were thinking that when we did not return to Spilsby at the time we should have, we may have been shot down. Later, when I did get back to Spilsby they were concerned that I was a goner and that they would have to get another pilot. It was an emotional welcome home they all gave me. Perhaps they had become used to me being with them. This confirmed I had a great crew. It was rewarding to realise that because of my always insisting on our crew remaining a tight unit, each and every one of us relying on all the others. At my age, then twenty years, I felt pride. I must now 'press on'...

CHAPTER 7: ON 'OPS'

Our first target with myself in charge was Harberg, as I recall, an outer suburb of Hamberg. Synthetic oil production was a large part of Germany's necessity. That was our target to destroy here. Our bomb load consisted of one 'cookie', about twice the length of a standard oil drum of that era, packed tight with high explosive material, plus other similar but lighter bombs, a sum total of 11,000 high explosives. Our bomb aimer (BA) released our load as he later told me, bang on target. At the moment of "bombs gone" we felt our Lanc, 11,000 pounds less weight, lift up as with relief.

We then turned to commence our return to base. There had been a lot of ack ack fire coming up at our bomber stream, also on the return journey, as well as night fighter activity, but our good fortune held. Our Lanc (A-Able) had avoided all that fire, ascending from the ground, plus night fighters.

After our successful landing, we taxied round to our own dispersal pad. As on every return, we were picked up by MT (Motor Transport Division) and taken to a building in which each crew sat together at one table where an officer attached to intelligence section de-briefed us by asking us a wide variety of questions. These ranged from any particular event any of us may have observed at any stage of the journey, or any close visuals of fighter activity, or of any of our own fellow aircraft being seen on fire or being shot down. Without going into detail, suffice to say, the pilot always instructed the navigator to enter any such occurrence in his log. All that was told to our interrogator was collated later at Intelligence Office, to ascertain if anything of value and/or importance had come to light. When debriefed, we went to eat an 'operational meal'. From there, to bed which, after six hours airborne, plus the time being debriefed, plus time eating, we all fell asleep when our heads hit the pillow.

Memory has just flooded back. After we parked A-Able, and set foot on terra firma once more we exhibited our joy. We had made our first 'op' OK. Our united crew teamship had proven itself. My own personal reaction was one of immense satisfaction in my own piloting. Yet, at the same time, I realised we still had much to do and accomplish. I told myself to keep my cool, not to get too cocky. All my crew had thanked me for getting to the target and bringing us all home safely. At age 20 years I did feel pride. Sleep that night was sweet.

Our next 'op' was one of the very few Bomber Command flew in daytime. The target was Duren which my memory at this moment cannot recall, other than that the target was flattened. What I do recall is the presence of our own aircraft in large numbers, some quite close. As our ops were almost all at night what we now saw in daylight was that we needed to be constantly aware, to keep our eyes and brains alert in our night ops, not just for enemy fighters, but also to avoid collisions with our own aircraft.

I shall now move on to our fifth op. It was one of the most unusual trips. The target was at the Norwegian port of Trondheim on the west coast. Norway had been attacked and defeated by the Nazis who now fully occupied and brutally controlled this country and its noble people.

Trondheim was a port about midway between its southern and northern tips of its North Sea coast line. It was used by the Nazis as a base for its U boats. They had built pens the U boats could sail into to refuel, also rearm with torpedos, as well as affording rest time for the crew members. Those pens had been constructed by long tunnels being basted out in the high hills above them. The entrance and exit from a waterway which was between the coast and the inner part of the land. This formed an isthmus. Along both shores of this waterway, the Germans had set smoke pots which, in the event of an air attack, would be activated so as to blot out any clear vision of the route into the entrance to the sub pens.

The plan devised at Bomber Command was based on the hope that our bomber force could take a route which could evade much of German radar and hopefully by going inland after crossing the coast at a point that was sparsely populated. This was advised by Norwegian resistance members who managed through secret radio contact with British radio bases. Thus, by flying inland for a specified time, being winter, the land was covered completely with snow. The stream would then fly north to a point equal to the target, which it was hoped, would enable us to get to the target in time to allow at least some of our planes to unload their cargo before Nazis started their smoke pots. At our lengthy briefing we were informed that there were many Norwegians living close to that waterway thus, if we were unable to drop our loads before the smoke pots were activated, we must abort the attack so there was to risk to our Norwegian friends.

The op commenced on Time. We flew at a low but safe height over the North Sea, crossing the Norwegian coast at the correct point, keeping out of range of the Nazi radar, flying inland to the point planned where we flew north to where we turned west towards the target.

The plan had been designed in the hope that the Germans would think that any attack would come from the sea towards Trondheim and not from inland. But the German were awake. As soon as the first two or three bombers started their approach bombing run the smoke pots were activated, completely blocking any vision. The bombing leader immediately radioed to all of us to abort the op and fly out to sea and when well away from the coat to jettison our bombs then return to our bases. Unfortunately the Germans had crews round the clock to quickly set the thick smoke clouds billowing out.

Had we succeeded to close the way into and out of those pens, they would not have any other port on the coast which could leave the U boats a safe haven. But in hindsight it was worth the attempt as it let the Germans know that we were a force not to be ignored.

When we landed back at Spilsby we had been airborne for eleven hours and thirty five minutes, during which I had been anchored in my seat. In all my life I have never had such a sore backside.

Our next op was a memorable one for its own reasons. The target was the famous city of Munich. At our crew briefing the intelligence officer informed us that at Munich, famous for its beer halls, weekend parties and which had not been attacked for quite a long time, we should catch them by surprise. We will be taking off on this Sunday night, and H hour, time over target, would be 3 am German time at which hour our leading squadrons would kick off the assault, dropping their 12,000 pound 'block busters', full of high explosives. But here, first let me describe our journey.

From take-off over France we flew at a low level, keeping below radar. Our course took us to a point at which we needed to climb to an altitude where we flew through the Swiss Alps, on the south side of Mont Blanc, Switzerland. Being a neutral country, its cities did not have a 'black out' as was the case in Britain, and it became a visual treat with the populated areas lit up. We all remarked on, and enjoyed, the sight of streets and lights. This was a scenic trip. All too soon we passed through the Alps, and the business at hand was resumed, being now over Germany.

After the short flight up to Munich we should surprise the Nazis. The briefing officer had predicted that the first anti-aircraft battery would take five minutes after the first bomb hit the ground before it fired off its first shell, because their crews would probably be the worse with booze. His prediction was spot on. Was this pure psychic?

We were at an altitude of about 18,000 feet when the first block buster hit the ground and exploded. A-Able shuddered, even at our altitude. Our cargo was only one 1,000 pound high explosive, the remainder being incendiaries, making sure that the target was set well on fire, the reason being that such high conflagrations would ensure German fire fighters faced an impossible task.

We had also been briefed that Munich was a centre of very much material war production. This attack was later declared to have a large effect on high loss to the Nazi of much badly needed war production output. Munich by moonlight, lit up by booze and bombs.

On return home, we had been airborne for 9 hours and 30 minutes. Once again, A-Able came through unscathed. She was the oldest kite in our squadron.

The next op: the dam on the river ERFT. Our 1000 pound bombs just seemed to bounce off the dam wall. Not a very good result over all. No damage to us.

Now, Munich again; but unfortunately A-Able's engines iced up so much that she was unable to reach the required height to get through the Alps, so we boomeranged. Maybe she was too old for mountain climbing. Next day, she went through a complete examination, a fault being discovered, then rectified, passing full return to operational flying.

Next op: Port of Gydinia. This turned out to be a difficult target: many German North Sea naval units in port, very heavy defences and much light flak on our bombing run, thankfully missing A-Able. Jerry ships noted trying to get out of port quickly. On return to Spilsby, due to large cloud cover we had to land at Kelston, and return to Spilsby the next day. We were the last of 44 Squadron aircraft, of those which had been diverted to Kelston, to return, becoming the last one to be refuelled and rearmed for an op to be flown this night. We were taken straight to the briefing hall.

The target was Politz, from memory, a north sea port of occupied Poland. At briefing, we were told that because so many aircraft returning from the previous night's diversion with hurried refuelling as well as engine tests, many crews would not be taking off at the time required to get to the target on the very last arranged time over the target. Therefore, if any crew was unable to reach target by that H hour they were ordered to abort the op and return to base.

A-Able, being the last kite to, at last, take off I knew the speed we would need to reach the target on time, as set at briefing. Hence, intercom talking with the navigator and myself, I decided to fly at an airspeed higher than our usual cruising level. It was a clear night, all the stars were visible, and the navigator spent a lot of time in his astrodome bubble in the top if the fuselage, using his sextant to get star shots, thereby getting accurate readings to get an accurate 'fix' on our position at each time. This enabled him to get his navigating spot on. Navigating was a special and Alf was a skilled navigator. He reported to me that we would reach the target if we could retain our present airspeed.

Good – oh! I thought. But about ten minutes later the starboard inner engine packed up. I had to shut it down. Thus on three engines only we would not be able to fly at the airspeed required.

At the briefing we had again been informed that the met office predicted that, at the time of our return to base, all England would be covered by thick fog. Thus, we would all be diverted to Dyce, an R A F station a little distance outside the northern city of Aberdeen in Scotland. That station was an operational training unit for pilots who were being trained to fly in Photographic Reconnaissance Unit (PRU). The aircraft they were flying was the famous De Havilland Mosquito, powered by two Rolls Royce Merlin engines.

One drawback for us was the short 1300 yard runway. Our shortest bomber runways were 1600 yards. Landing with only three engines required a higher approach speed hence I was to alter our normal procedure. This would require my judging to set A-Able to touch down on the very beginning of that short concrete runway, then gently braking earlier than normally. Also, the in to wind speed was quite low, meaning I would need to start braking harder and sooner than on a normal landing. My concentration was at its peak. I succeeded in bringing A-Able to a stop at the very last foot of the concrete. You can understand, dear reader, my intense sigh of relief. Cyril, our flight engineer tapped me on the shoulder saying on the intercom, "Great show Skip", which was echoed by the rest of the crew.

When we had been guided by a ground crew bod to a nearby dispersal pad, I sat still for a couple of minutes, just thinking, before I managed undo all my chute harness and other safety gear, then struggled to make my way down to the exit door. All the crew were waiting outside. As I got to my feet on the ground I found myself being hoisted on to shoulders with voices singing "For he's a jolly good fellow". I felt a tear or two trickle down, as indeed I feel at this moment, in memory even now shedding tears. What a terrific crew I had.

That highly intense mental and physical effort drained me. It was quite a while before I regained normality. It also, I realised, must have filled each crew member, because all of us, with our operational flying so far, were well aware of the hazards that landing was going to entail. Their combined relief most certainly would have equalled my own. However, I slowly felt, once more, a measure of pride in our whole crew effort.

This had occurred just before Christmas 1944. When the engine failure had been diagnosed, and engine crew needed to come up from England, along with necessary spare parts for replacement, hence we remained in Dyce until January 1.

Therefore, I experienced that great annual celebration of 'Hogmanay". (I do hope I got that spelling accurately, if not, I do apologise profusely to my Scottish ancestors).This celebrates the coming in of each new year on New Years' Eve, which this year I spent in the Officers' Mess. I did not have any money with me. I cannot recall how I managed at the end of that night, to finish up being 'foo the noo'. If you are unable to understand that phrase, find a Scot who will translate it into English.

Recall that, at the briefing for this op, we were briefed to divert to Dyce because fog blanketed all of England. After we had landed at Dyce the wireless signal went out to all aircraft that the forecast fog had cleared, thus all could now return to Spisby.

However, another 44 Squadron kite (undamaged) Y-Yoke landed at Dyce, captained by F/L Les Hayler, one of those very fine Rhodesians. His aircraft had actually reached the target and bombed on time. The wireless set in Y-Yoke had developed a fault, to the extent that the signal to all aircraft not to divert to Dyce, but to return to Spilsby, was not received by Y-Yoke's wireless operator, thus they went to Dyce as per briefing.

The next day, Les said to me (still a little the worse for wear for New Years' Eve) that he was going to take Y-Yoke up for a flight test and would I like to come with him for a ride. I agreed. We were taken to his kite by an MT vehicle which was a normal flat deck truck on which was placed a separate canopy. Inside it seating was set it and there were big areas of glass along each side giving a clear view of the outside. There was a lot of un-mowed grass alongside that one runway.

After our brief air test and landing (I was glad I was not the one who had been driving), we were picked up by the same MT vehicle. On board were a few other aircrew bods from other aircraft. About two minutes after pick up an obviously Cockney voice rang out loudly, "Look at this bleedin' 'Spit' (Spitfire) comin' in to land", and there's an aircraft on the runway about to take off. All our eyes immediately turned to the view of a single engine fighter heading towards the runway as if to land on it. At the last minute, the pilot must have become aware of the kite starting its take off run.

The incoming pilot suddenly gunned his engine, flying away to one side. All this time, the Cockney voice kept up a running commentary of the entire circus. The Spit completed a rapid circuit coming in to a good landing on the long grass, slowed quickly then lurched over forwards, ending up on its back.

The driver of our MT vehicle stopped, rushing round to the rear, opening our canopy doors, from which we all alighted, rushing to the overturned aircraft. A very 'posh' English voice of authority called out loudly, "The pilot is trapped upside down in his cockpit. All personnel spread along under both wings, your backs touching under the wings, then on my order 'lift' raise your bodies in unison". He gave the order and we all lifted together. The 'posh' voice rang out, "Good show men, he (the pilot) is almost clear enough".

At that moment the Cockney voice yelled out "Gawd Blimey, this ain't a Spit, it's a bloody Jerry". On that, all we lifters lowered our bodies, the kite going down again. 'Posh' again took over, "So it's a Hun, get lifting again and no slacking". The exercise recommenced successfully to allow the pilot a safe exit.

When he was first set on his feet the poor obviously confused very, very, young German, was surrounded by us, his enemy. He was clad in a soft high quality one piece leather flying suit, with a built in watch on one wrist. He also had two pistols each in its own holster. A flurry of would-be souvenir hunters descended upon him, but only the pistols left his frame, one of the which the holster was also removed. The other one, the holster was left on his body. There was nothing else able to be taken. The poor bloke was quickly raced doff to be interviewed by the intelligence officers.

It appears that the Luftwaffe squadron in which he was serving had been based in Denmark but due to the western combined armies now starting to advance more rapidly towards Germany, that squadron had been ordered to deploy to a base in occupied Norway. This pilot disclosed that he realised that Germany, Hitler and all were doomed to defeat.

Thus, while his squadron left Denmark, flying to Norway, he stationed his aircraft at the rear of its formation and after Denmark had disappeared behind him, he called his leader saying he was having engine trouble and that he had doubts that he would be able to reach their destination at all. His leader ordered him to return to Denmark to have the engine examined. He acknowledged the order, turned about, flying slowly to conserve fuel, then when his fellow flyers had flown out of his view, he headed to the west where he planned to land and surrender at the first aerodrome he saw.

I presume that when he was taken into 'friendly' custody, there to be grilled further to try and discover any other info not yet already known. Putting myself in his situation, I am convinced I should cooperate fully.

One obvious fact which was known was that a pistol had been removed from its holster. The Tannoy (a loud speaker system installed around each RAF station) blared out, "Will the person (singular) who removed the pistol from the captured German airman surrender it immediately to station headquarters. If not, measures will be put in place to recover it". I did not know what the outcome to that demand was.

No mention was made about the other pistol, which had been removed with its holster as well. I would consider that, in the heat and scramble of the rush of would-be souvenir hunters at the time of the incident, nobody on a position of authority, even 'posh' had observed that the German had been wearing two holstered pistols.

I seem to recall that rumour had it that a very quick-thinking Australian had employed his intelligence to bring about a crafty vanishing act. Some win; some do not.

What a multi eventful New Year interval this turned out to have been. Yes, dear reader, all this DID occur.

A-Able was fixed and we returned to Spilsby. Back at Spilsby, first 4 days flying on training, fighter affiliation, in which a fighter aircraft was towing a drogue, which was like a windsock. The gunners shot at the drogue, not with ammunition, but recording cameras testing their deflection aiming. Both rear and mid upper gunners had good results.

Next operation, January 5th 1945: the target was Royan, on the west coast of France. That section was Vichy France, that part of the country which the Germans, in their great generosity, had allowed the French to rule for themselves, yet still with cunning, the Huns oversaw. At the crew briefing we were informed that a section of Yankee bombers would also be joining us to bomb. This created a loud laughing amongst all of us; Yankees flying at night? They could never navigate at night that is why their formations flew only in daylight hours. We are still waiting for them to arrive.

Next night, January 6th: target Houfalize. This was during the German counter attack in the Ardennes forest by German General von Rundstedt which became known as 'The battle of the Bulge', because his tanks had made an advance in the allied armies' line. Why did they not get further? Their tanks ran out of supply of their diesel fuel. Why? Simple: the RAF strategic bomber offensive, during which oil targets were constantly attacked. Large snowfalls had unfortunately blotted out target indicators badly. However, we afterwards learned that large numbers of German troops were 'eliminated'. On return, we were diverted because of fog at Spilsby, to Croft in Yorkshire. Then back to Spilsby, We were on ops in the English winter, hence frequent diversions.

January 14th Target Mersberg, 10 x 1,000 lb bombs. A-Able copped some hits from light anti-aircraft fire. Luckily no crew member was injured.

January 16th Target Brux in East Germany; bomb load, 1 x 4,000 lb high explosive (a cookie), plus many incendiaries. A bit more flak damage, also rear gunner sighted a Focke-Wulf 190 fighter coming at us from the starboard quarter. The gunner opened fire at it. Joe thought he might have scored a hit, however it was not sighted again. No crew hit. We returned safely, but things seemed to be hotting up. We had a short break before our next op.

February 1 : Target Siegen, load 12 x 1,000lb bomb.

February 2 - 3: Target Kahlsruhe, bombed well. We endured two fighter combats, the first attack from an M/E 110, the second by a JU 88. We successfully corkscrewed away from both. The rear and mid-upper gunners got some shots fired off on both occasions without knowing if they scored any hits.

February 7 – 8: attack on Dortmund Emms canal. The Huns were manufacturing submarines in sections in inland places, then sending these sections on barges down canals to ports for final assembly at the port area. The intense naval anti-U boat offensive had decreased the numbers of U boats at sea by a considerable amount. This had the follow-on effect that many more merchant ships bringing vital supplies to the UK arrived safely, hence the air attacks on the barges in the canals was creating a double-edged sword by reducing the number of new boats getting to sea, as well as, in many instances, damaging the canals, slowing their traffic to ports. Our bomb load was 12 x 1,000 lb high explosives. This attack put a large dent in German war production.

February 4th (my 21st birthday) we were training it was for high level bombing, of which we had already done a lot of ops, but all crew were still detailed to carry out training exercises. On this particular day we had the honour of being accompanied by the station commander, Group Captain Harris. I do not know whether he had personally asked to fly with our crew, or if it was a custom to fly with a crew, which was then decided by the station commander. Whatever, after we had landed, he personally commended me and our crew saying "Keep it up MR Walker". That had been a one hour trip. The 'Groupy' was a popular and true gentleman. All station commanders were former pilots, some from World War I.

February 13/14 - The Target Dresden. This was the furthest East German distance we had flown.

Starting out I am going into a lot of discussion and explanations because much publicity and false statements were subsequently made about it by grossly ignorant people.

At our crew briefing we were firstly addressed by a senior intelligence officer from Bomber Command headquarters at High Wycombe. He commenced with a full explanation of the state of the overall war situation existing at this time.

The western armies, British, Canadian and Yanks, were now making faster advances against the German troops on the western front. From the east, Russian armies were led by the biggest tanks of any of the countries involved. They created absolute havoc to the returning tanks for the retreating Huns. That retreat was hampered by the thousands of German civilians trying to escape from the Russians, cluttering the roads whereby German troops retreating were experiencing traffic blockages.

Dresden city, consisted of mainly timber buildings, built centuries earlier. The city was NOT our target. The eastern sector of the German railway system, not ever having been attacked, was still in full operation. The Target, the very huge railway marshalling yards at this time were full of rolling stock which had been, and was still being loaded with war supplies to be sent to both their western and eastern war fronts. Of these, tank fuel, armaments, food and ammunition were priorities.

Our task was to completely destroy those marshalling yards. Our pathfinder aircraft would lead, dropping illuminating flares, then target indicators, on which bomb aimers were to focus their bomb sights.

We (our crew) were one of 250 Lancasters of 5 Group to form the first wave to attack that target. A short spell later, a larger force, guided to the target by the massive fires already enveloping the area, added their cargo. Those yards were utterly destroyed. The huge fires burned up all the oxygen in the area, which in turn created gale force winds to replace the oxygen thereby taking fire to the adjacent city buildings. That city, then in turn, suffered huge destruction, incurring much loss of life.

The main object of that op was that it fully destroyed vital supplies for German forces, thus completely denying them. From the information given us at the briefing, my own opinion is that one op caused a shortening of World War II.

The aftermath brought a report from a self-proclaimed 'historian', David Irvine, claiming that 60,000 civilians were killed. Certain British clerics were agitating to have the RAF bombing halted completely. The most prominent of those clerics was the Bishop of Chichester. The thought has gone through my mind that the character in earlier history, Mr Spooner would have referred to him and the Chishop of Bitchchester.

Oddly enough, in an article I read somewhere about two years ago, that today's Dresden local body people discovered an underground cellar-like space, in which they discovered archival records of Dresden. One of those was a record of population in the year referred to which was then, by extension, calculated to a figure probable in 1945, to a figure of likely 30,000 inhabitants less killed than claimed by Irvine. Q.E.D., Quad Erat Demonstrandum, Latin for 'Which was to have been proved' – verbal amplification maybe. On reflection the large numbers of civilian houses I had seen flattened, with thousands killed in the UK, our own much heavier bombing was more than justified. The Cishop of Bitchchester made no mention of Britons killed.

Next op: the night after Dresden; target Rositz, cargo, 1 cookie plus 9 x 500 lb bombs. On the way we passed Chemnitz which was in the act of being 'pranged' by the rest of Bomber Command. We, the squadrons of 5 Group were the only group on Rositz. Getting close to target, received some few hits by light flak; no crew hit. This was followed by a running combat with a Jerry night fighter. Fortunately, our mid-upper gunner spotted it early enough, so that I was able to evade the Jerry, which was lost to our sight in intermittent cloud, which in itself made the bomb aimer's task quite a sticky one. Nevertheless, he did get his bombs away.

Five nights later, target Bohlen, with thick cloud over the target, causing somewhat of a shambles. Thus, the photographs taken of bombs when hitting target showed only cloud. I seem to recall that in this op the Pathfinder bombing controller (called master bomber) failed to return.

The following op, four nights later, target Mittelland canal. Our cargo was 12 x 1,000 lb heavy explosive bombs, each with ½ hour delay fuses. Maybe that was to explode when German repair crews arrived, thus being 'eliminated'. Jerry fighters reaped a harvest of Lancasters; I saw one crash on a German airfield. However, we brought about huge damage.

Five days later, the Squadron instructor F/L Webster, still a grim sort of bloke, took us up to carry out a 10/20 'sortie check' which we must have passed, without being told the result. I suppose by now we were classed as an 'experienced crew' (fingers crossed).

Our next op: Sassnitz, a port on the North Sea coast of Germany. This was 'put on' in a hurry, because intelligence had received reports that quite a number of German warships of their North Sea naval fleet had arrived and berthed there. We flew low over the North Sea to evade German radar then climbed to our bombing height. Pathfinders dropped their illuminating flares, then their target markers, while we, A-Able, were just starting our run in, quite a fair way back from the target, when I saw a big orange and red ball explosion ahead of us. I commented on the intercom to our crew, "Some poor bods have been blown to smithereens ahead of us". Then back to our own bombing run. Our bomb aimer got a good sighting of the target indicator, after which we scurried away as soon as possible.

In the early stages of our return over the North Sea many German fighters were chasing us. We were attacked twice, but our two wonderful wide awake gunners saw the Jerries early, resulting in my getting corkscrew flying in time to frustrate the 'Bosch' pilots who were unable to get a 'bead' on us.

When we landed I was extremely fatigued, thus at debriefing I did not look at the blackboard listing the aircraft which had safely returned. The next day we learned that our attack had been highly successful and also we had suffered only one aircraft which had not returned.

I was completely disturbed when I learned that my very best mate Buzz, a Canadian, and his crew were the bods who did not return. His was the Lanc. which I had observed being blown up in front of us. They could only have had a direct hit by a heavy high explosive anti-aircraft shell. With that sudden hit, poor Buzz would not have known what hit them. They would all have been killed instantly. Vale Buzz; I still often think about you, even well over seventy years later.

Fortune came my way. I was given a ten day leave pass. For that I travelled in the famous train, the 'Flying Scotsman', to Edinburgh. The name of my home town in New Zealand, Dunedin, I learned was the Gaelic for Edinburgh. That, thereby, added to my vast store of completely useless information. However, I did get to an international services rugby match between England and Scotland. Within both teams there were about six NZ All Blacks, some pre-war, some post-war, also two Aussie Wallabies and one South African Springbok. It was played on 17 March 1945 (Saint Paddy's Day) but I can't remember who won. A great camaraderie among the players, Empire teamwork on the playing field, as it was in all our wartime flying. Oh where have those great ties gone in these times.

March 20/21 target Bohlen again: cargo 1 cookie plus 9 x 500lb high explosives. Much flak about, missed a collision over target by a hairs' breadth; a disturbing event. But that intense training before going on to an operational squadron meant all crew were always keeping a lookout. Our mid-upper gunner this time gave me the near collision warning. Thanks Jock.

Our last op, Hamburg: 1 cookie and 11 x 500lb high explosives. Starboard engine on fire, after fighter fired from underneath us. Extinguished by operating Graviner switch, sending flaming extinguisher chemical through perforated piping, surrounding each engine. Flight engineer, young Cyril, performed that exercise promptly. Again, each crew being important. Returned on 3 engines.

End of operational flying. We had survived! Fine celebrations; each one of us sensing great relief. German last ditch defensive ability almost, at this time, unable to resist its total destruction.

However, a big blip appeared on my horizon. War in Europe almost over BUT war in the Pacific was still in full flight.

We Kiwis were still under the authority of the RNZAF headquarters in the UK. I received orders from that HQ to immediately report to NZ section in Brighton, which, reader, you will recall was formally our receiving centre when we first arrived in the UK after our aircrew graduating. Now it was the departure centre from which we were returning to NZ.

I was directed to report to a squadron leader whose name long ago vanished into this air. He told me that in NZ I would have a different role to play.

Memorial to those who served in 44 (Rhodesia) Squadron Royal Air Force.

I embarked on a troopship leaving Liverpool in July. Aboard there was a contingent of former Kiwi army men who had spent years as prisoners of war in Germany. One of them I was talking with I recognised as a classmate at High Street Primary School in the 1930s, whose name I recall was Bill Gardner. I said to him "Hello Bill". I was, of course, still in my officers' uniform. He replied "Who are you sir?" I said, "Forget the sir, I'm Ken Walker, we were at High Street school together, we are mates". Bill then remembered me. Being introduced to the pals he was with, we all enjoyed a fine time together. Rank was forgotten.

During the journey home we passed through the Panama Canal. While we moved through the Canal the ship's wireless lost reception from all countries' broadcast news. On entering the Pacific, the new was again able to be picked up. The startling news was that the Yanks had dropped the first atom bomb on Hiroshima. Two days later the second A-bomb had hit Nagasaki. The Japs would not comply with the American demand for unconditional surrender, but some days later, when they were firmly advised that if they failed to completely surrender, further A-bombs would rain down on Japan. That did bring about their full surrender. Therefore, on August 14th, the war in Europe and Japan were both over.

Hence, when our ship berthed in Wellington, it was a short walk with my kitbag to RNZAF HQ in Short Street.

On reporting, as briefed back in Brighton, to the officer named, he informed me that the intention had been to have me converted to a single engine NZ squadron operating in the Pacific, but now that both wars had concluded, the NZ government had made it clear that they had no intention to retain the Air Force on a war footing. Thus, I was offered an immediate discharge, which I accepted. I was given travel vouchers back to Dunedin where I was greeted at the railway station by my brother Jack who had also seen action in the army in parts of the Pacific. He had also been demobbed, and both of us were now in 'Civvy Street'. On reflection (now 2018), it was something of a shock for quite a long time.

I had left New Zealand as a 19 year old boy and, on return I was a 21 year old man. That spell in what was the longest continuing battle of that recent time 'The Battle for Germany' had taken the greatest toll in lives, mental and physical disabilities, among all of the sectors in Western allied forces. At this very moment, while nearing mental exhaustion in the penning of my journey, be patent dear reader, but I must take rest, but I shall return.

Now I have somewhat sufficiently recovered, my Air Force journey must end here.

I thank you, dear reader(s) for your patience and indulgence. Cheers, Ken.

HIGHLIGHTS IN MY JOURNEYING

- My first solo in the Tiger Moth.

  At that time the top song on the Hit Parade was called 'The Tiger Rag". I remembered one line was 'hold that Tiger'. Thus, on the upwind leg of my solo I found myself at ease, confident in my new found freedom, and I started singing to myself, "Hold that Tiger" over and over. On landing, I did a perfect three pointer; that is when both wheels and the tail skid touch down at the exact same time. I taxied over to the edge of that grass field where P/O Paynter had got out, telling me, "It's all yours now", and where he was waiting for me. Sayeth he: "Walker your landing was a perfect three pointer and probably you will never do another one. His prediction was spot on.

- Passing the Course.

  Being notified that I was one of those who passed and was being posted to Canada to the next training course, SFTS. A few days leave then up to Auckland, boarding the troopship Matsonia, never having done sea travel. It seemed so exciting.

- That wonderful unforgettable train journey over the Canadian Rockies; all for free. At age 18 this was hard to believe.

- Getting to train on a twin engined aircraft.
- This was a step up. Passing through the course.
- Getting my pilot's wings presented with the rank of Sergeant.

  If only my dearly loved Dad had not died in 1935, he would have been so proud of me. Dad, I still love and miss you.
- Boarding and sailing on the famous Queen Mary.
- Next high spot: Actually (these days mostly mispronounced as 'akchilly)
- Then on arrival off England's shores, being welcomed by two Spitfires giving us a great flying display.

  At last, treading on English soil: this is where a multitude of events during centuries past have been played out. A few days leave, then going up to London. Viewing many places and buildings of renown. This was akin to turning the living pages of the history books we had swotted in our school days.

  Highlights from then onwards were really absorbed in our progressive serving in the training which, was indicated during my journey.

- Converting on to the Lancaster

  This was the next special moment. The Lanc (as it was affectionately referred to) had already become the icon in the bombing family. Then:

- Arriving at Spilsby, becoming part of 44 (Rhodesia) Squadron.

  The Latin motto on the Squadron crest 'Fulmina Regis Iusta': 'The King's thunderbolts are just'.

  From then on, each page in itself was a highlight. But our first op with myself in charge, until our last op, which, to our joy, "We had survived!" We were alive. We had become experienced crew. That, in itself, was surely the supreme highlight.

THE AFTERMATH

We won that war – BUT in my own opinion – we have lost the peace. Why do we not, as ordinary people ask ourselves, WHY?

At the end of WWI the Treaty of Versailles was held, in which the politicians were to decide what punishment and/or strictures were to be placed on Germany.

At that Treaty, the UK was represented by its Prime Minister, David Lloyd George. It was recorded that when he returned to the UK, he was asked the question: "How did things go at the Treaty meeting?" His reply was, in this view, "Oh we talked a lot amongst the leaders of our allied countries, and the bankers arrived, virtually taking over".

Now, in the year 2018, have we really learned. Bankers worldwide seem to be in control. I try to think for myself, and not believe the would-be 'know all's' in the various media. I have worked out that our media are mostly in the ownership of giant world corporations. The most influential person in each form it takes is never the editor it is the headline editor e.g. 'PM slams statement made by shadow energy minister' all in heavy black type. The majority of people do not read, listen to or watch the content coming past the headline. Do we ask, "How does money come into existence, in how many different forms does it exist?”

Yes we did win that war, but puzzle to find the peace. As world life, also Australian life is not what we believed we were fighting for, failed to justify our wartime beliefs. Ask yourself - WHY? There are multitudes of WHYs.

I have not, these days, long to live, Hence the self-questioning must come, else in what state will humanity exist. Winning a war does not prove who is right, only who is left.

Dear reader, please ask yourself many questions like “WHY”. Why is this planet seemingly close to self-implosion? I am now too old to lead, or be part of the solution.

SOME TERMS IN COMMON AIR FORCE JARGON DURING WWII

- An aircraft – a kite
- Motor Transport Section – MT
- Parachute – a chute
- Women's Auxiliary Air Force – WAAFS
- New recruits who as yet know nothing – Sprogs
- Lower rank personnel, always ground staff – erks
- Wing Commander – Wingco
- Group Commander – Groupie
- Girls – Popsies (I particularly like that word)
- Persons – Bods (short for bodies)
- Aerodrome – drome

- Dining room – Mess There was one for the commissioned ranks, a different one for the 'other ranks'. One could imagine that in these modern-days' political correctness, a host of protesters howling "discrimination". My comment: I am so glad that I lived and served in those days when 'common sense' still prevailed. We were not swathed in cotton wool.

- Bombing operation – op

- A 4000 lb high explosive bomb – a cookie

- My only decoration was the MHDOIF – The most highly derogatory order of the irremovable finger.

- New Zealand's emblem was the kiwi, a flightless bird, yet many Kiwis did fly.

ROYAL N.Z. AIR FORCE
PILOT'S FLYING
LOG-BOOK
N.Z. 421685
F/O. K. H. WALKER D.F.C.

INSTRUCTIONS

[see K.R. & A.C.I., para. 786.]

1. This book is an official document and is the property of His Majesty's Government.

2. An accurate and detailed record is to be kept in the log of all flights undertaken in service aircraft by the individual to whom it relates.

3. Monthly flying will be analysed by aircraft types and inserted in red ink. The stamp will be inserted on the left-hand page appropriately aligned to the ruling.

4. The annual summary and assessment will be completed on Form 414A and inserted in the appropriate page of the log. This form will also be used when a pilot is posted or attached to another unit for flying duties.

NIGHT VISUAL CAPACITY

Exceptional
Above Average
Average
Below Average
Poor
Score 27

Date 2-10-43

CERTIFICATES OF QUALIFICATION AS FIRST PILOT.

[K.R. & A.C.I., para. 805.]

NAME: WALKER, KENNETH HOLLWAY. RANK: LAC SGT F/SGT. P/O F/O

(i) Certified that the above-named has qualified as a first pilot (day) on T. E. landplanes w.e.f. 17. 9. 43.

Unit: 15 SFTS - RCAF.

Signature:

Date:

Rank:

(ii) Certified that the above-named has qualified as a first pilot (night) on .... landplanes w.e.f. ....

Unit:

Signature:

Date:

Rank:

(iii) Certified that the above-named has qualified as a first pilot (day) on .... seaplanes w.e.f. ....

Unit:

Signature:

Date:

Rank:

(iv) Certified that the above-named has qualified as a first pilot (night) on .... seaplanes w.e.f. ....

Unit:

Signature:

Date:

Rank:

H1 44 (RHODESIA)

YEAR: 1944 MONTH.	DATE.	AIRCRAFT. Type.	No.	PILOT, OR 1ST PILOT.	2ND PILOT, PUPIL, OR PASSENGER.	DUTY (INCLUDING RESULTS AND REMARKS).
—	—	—	—	—	—	— TOTALS BROUGHT FORWARD
NOVEMBER	3	LANCASTER	"R" 417	SELF	CREW	EX 3 - D.N.C.O.
	4	"	"L" 654	SELF	CREW	EX 3 - X COUNTRY.
	5	"	"D" 991	SELF	CREW	- X. COUNTRY
	6	"	"L" 654	F/L DAVEY	SELF, CREW	1ST OPERATION - MITTELLAND CANAL, - GRAVENHORST - 13 x 1000 lb - ABORTIVE
	7	"	"B" 631	F/O DAGGARTT	CREW	WOODBRIDGE - SPILSBY
	9	"	"U" 655	SELF	F/O MARSH, CREW	EX 1. - 3 & 2 ENG. FLYING.
	10	"	"Z" 372	F/LT WEBSTER	SELF, CREW	EX 2. F/A, 3 ENG. FLYING
	11	"	"D" 991	SELF	CREW SGT CROOKS " TURNER " HARDY " ROBERTS " SPENCE " WARD	HARBURG - 2ND OPERATION 1 x 4000 lb. 6 x 1000 lb 6 x 500 lb
	14	"	"H" 625	SELF	CREW.	H_2S - P.P.I. RUNS.
	16	"	"A" 496	SELF	CREW	3RD OPERATION - DUREN 11 x 1000 lb., 4 x 500 lbs.
	21	"	A 496	SELF	CREW	4TH OPERATION - MITTELLAND CANAL 12 x 1000 lbs.
	22/23	"	A 496	SELF	CREW	5TH OPERATION - TRONDHEIM 6 x 1000 lbs. 4 x 500 lbs - ABORTIVE
	26/27	"	A 496	SELF	CREW	6TH OPERATION - MUNICH 1 x 1000 lb 13 J TYPE CLUSTERS. - WIZARD TRIP - Jenny caught unawares

GRAND TOTAL [Cols. (1) to (10)].
509 Hrs. 00 Mins.

TOTALS CARRIED FORWARD

DAY.		NIGHT.		DAY.			NIGHT.			PASSENGER.	Cols. (1) to (10)].	
DUAL.	PILOT.	DUAL.	PILOT.	DUAL.	1ST PILOT.	2ND PILOT.	DUAL.	1ST PILOT.	2ND PILOT.		DUAL.	PILOT.
(1)	(2)	(3)	(4)	(5)	(6)	(7)	(8)	(9)	(10)	(11)	(12)	(13)
38.45	26.55	4.00	-	139.40	137.25	2.10	38.00	62.55	-	67.25	63.30	20
								.55				
								4.20				
								4.35				
								-	-			
SHOT UP BY CONVOY RECROSSING ENGLISH COAST. HYDRAULICS SHOT AWAY. LANDED AT WOODBRIDGE EMERGENCY LANDING FIELD.									6.00			
										.40		
					.50							
				1.40								
BAGS OF HEAVY FLAK. MILLED ABOUT OVER TARGET.				-								
VERY HEAVY ICING IN CLOUDS.					-							
						-						
							-	5.40				
					1.35							
SKY BLACK WITH AIR/C. TARGET FLATTENED					6.10							
HOLED BY FLAK. VERY CLOSE SHAVE FROM A COLLISION WITH ANOTHER LANCASTER.												
								6.20				
[illegible]0 gallons of juice left. I never had such a close behind in all my life. Scenery over Norway lovely								11.35				
								9.30				
Trip over the Swiss Alps in moonlight was the most beautiful sight I have ever seen.												
38.45	26.55	4.00	-	141.20	146.00	2.10	38.00	105.50	6.00	68.05	63.30	31

YEAR: 1944 MONTH. | DATE.	AIRCRAFT. Type.	No.	PILOT, OR 1ST PILOT.	2ND PILOT, PUPIL, OR PASSENGER.	DUTY (INCLUDING RESULTS AND REMARKS).
— —	—	—	—	—	— TOTALS BROUGHT FORWARD

SUMMARY FOR NOVEMBER 1944

		DAY	NIGHT
OPERATIONS	DAY	6 · 10	·
OPERATIONS	NIGHT		39 · 05
TRAINING	DAY	4 · 05	
TRAINING	NIGHT		9 · 50
	TOTALS	10 · 15	48 · 55

DATE 10 : 12 : 44

SIG. K. Walker

SIGNED .... S/L O/C "A" FLT

SIGNED .... W/CDR O/C. 44 SQN.

GRAND TOTAL [Cols. (1) to (10)].
509 Hrs. ✓ Mins.

TOTALS CARRIED FORWARD

Single-Engine Aircraft.				Multi-Engine Aircraft.						Passenger.	Instr/Cloud Flying [Incl. in Cols. (1) to (10)].	
Day.		Night.		Day.			Night.					
Dual.	Pilot.	Dual.	Pilot.	Dual.	1st Pilot.	2nd Pilot.	Dual.	1st Pilot.	2nd Pilot.		Dual.	Pilot.
(1)	(2)	(3)	(4)	(5)	(6)	(7)	(8)	(9)	(10)	(11)	(12)	(13)
38.45	26.55	4.00	-	144.20 ~~[illegible]~~	146.00	2.10	38.00	105.50	6.00	68.05	63.30	81

YEAR: 1944 MONTH.	DATE.	AIRCRAFT. Type.	No.	PILOT, OR 1ST PILOT.	2ND PILOT, PUPIL, OR PASSENGER.	DUTY (INCLUDING RESULTS AND REMARKS).
—	—	—	—	—	—	— TOTALS BROUGHT FORWARD
DECEMBER	10	LANCASTER	"A" 496	SELF	CREW	OPERATIONS — RECALLED
	11	"	A 496	SELF	CREW	7TH OPERATION - RIVER ERFT DAM 14 x 1000 lbs.
	17	"	"A" 496	SELF	CREW	OPERATIONS - MUNICH / "BOOMERANG" WING ICING
	18	"	"A" 496	SELF	CREW	8TH OPERATION - GDYNIA [illegible] x 1000 lbs
	21	OXFORD	R 6430	F/LT RICHARDSON	SELF	SPILSBY - KELSTON
	21	LANCASTER	A "496"	SELF	CREW	KELSTON - SPILSBY.
	21	"	"A" 496	SELF	CREW	OPERATIONS . PÖLITZ / "BOOMERANG" - SI. CUT
	22	"	"Y"	F/LT HAYLER	SELF CREW	AIR TEST AT DYCE - WE HAD JUST FINISHED THIS &
	28	"	"Y"	F/LT HAYLER	SELF CREW	DYCE - SPILSBY

SUMMERY FOR DECEMBER 1944

		DAY	NIGHT
OPERATIONS	DAY	5 · 45	
OPERATIONS	NIGHT		24 · 10
TRAINING	DAY	3 · 05	
TRAINING	NIGHT		—
	TOTALS	8 · 50	24 · 10

[signature] .... S/L O/C "A" FLT.

[signature] .... W/CDR O.C. 44 SQDN.

GRAND TOTAL [Cols. (1) to (10)].
542 Hrs. — Mins.

TOTALS CARRIED FORWARD

DAY.		NIGHT.		DAY.			NIGHT.			PASSENGER.	FLYING [Incl. in Cols. (1) to (10)].	
DUAL.	PILOT.	DUAL.	PILOT.	DUAL.	1ST PILOT.	2ND PILOT.	DUAL.	1ST PILOT.	2ND PILOT.		DUAL.	PILOT.
(1)	(2)	(3)	(4)	(5)	(6)	(7)	(8)	(9)	(10)	(11)	(12)	(13)
38·45	26·55	4·00	-	141·20	146·00	2·10	38·00	105·50	6·00	68·05	63·30	31
								2·55				
					5·45							
								7·15				
								10·45				
						·30						
					·20							
								3·15				
						·30						
						1·45						
38·45	26·55	4·00	.	141·20	152·05	4·55	38·00	130·00	6·00	68·05	63·30	37

A SORE THORN IN OUR SIDE ALL THIS WEEK. 1000 lb ers JUST BOUNCE OFF THE DAM.

SWISS ALPS IN ICING WAS JUST TOO MUCH FOR OLD "ABLE" TONIGHT. SHE IS GETTING TOO OLD.

A VERY DIFFICULT Δ. ALL GERMANY'S NAVY IN GDYNIA HARBOUR & THUS DEFENCES VERY THICK & HEAVY & BIG LOSSES SUSTAINED. ENDLESS STREAMS OF LIGHT FLAK TRACER ON BOMBING RUN. - BOY DID THOSE JERRY SHIPS START SCATTERING OUT OF PORT!

LANDED AT "DYCE" - ABERDEEN ON 3 ENGINES. STAYED THERE FOR A WEEK! HAD A "BANG-ON" XMAS.

A JERRY SGT-PILOT WHO WAS CHEESED-OFF WITH THE LUFTWAFFE & BROUGHT A ME 109G IN & SURRENDERED HIMSELF. HE TURNED THE AIR/C ON ITS BACK WHEN HE LANDED.

YEAR: 1945 MONTH.	DATE.	AIRCRAFT. Type.	No.	PILOT, OR 1ST PILOT.	2ND PILOT, PUPIL, OR PASSENGER.	DUTY (INCLUDING RESULTS AND REMARKS).
—	—	—	—	—	—	— TOTALS BROUGHT FORWARD
JANUARY	2	LANCASTER	"D" 991	SELF	CREW	FIGHTER AFFILIATION.
	3	"	"Q"	SELF	CREW	FIGHTER AFFILIATION
	4	"	"L" 654	SELF	CREW	P. P. I.
	5	"	"A" 496	SELF	CREW	9TH OPERATION – ROYAN – ABORTIVE
	6	"	"A" 496	SELF	CREW	10TH OPERATION – HOUFFALIZE – ABORTIVE
	6	"	"A" 496	SELF	CREW	CROFT — BASE – 5 GROUP LINE SHOOT TAKE OFF
	14	"	"Z" 372	SELF	CREW	11TH OPERATION – MERSEBERG 1X4000 ♀
	15	"	"Z" 372	SELF	CREW	LECKENFIELD – BASE DID A DIRTY BIG SHOOT UP RIGHT ON THE DECK
	16	"	"L" 654	SELF	CREW	12TH OPERATION – BRUX – 1X4000 ♀

SUMMARY FOR JANUARY 1945.

OPERATIONS	NIGHT		32 · 15
TRAINING	DAY	5 · 40	
		5 · 40	32 · 15

H. F. Richardson F/L . . . S/L O.C. "A" FLIGHT

[signature] . . . . W/CDR O.C. 44 SQDN.

GRAND TOTAL [Cols. (1) to (10)]. 579 Hrs. 55 Mins.

TOTALS CARRIED FORWARD

DAY.		NIGHT.		DAY.			NIGHT.			PASSENGER.	FLYING [Incl. in Cols. (1) to (10)].	
DUAL.	PILOT.	DUAL.	PILOT.	DUAL.	1ST PILOT.	2ND PILOT.	DUAL.	1ST PILOT.	2ND PILOT.		DUAL.	PILOT.
(1)	(2)	(3)	(4)	(5)	(6)	(7)	(8)	(9)	(10)	(11)	(12)	(13)
38·45	26·55	4·00	-	141·20	152·05	4·55	38·00	130·00	6·00	68·05	63·30	37
	"				1·30							
					1·45							
					1·15							
JACK HARDY BOOBED WITH HIS SELECTOR SWITCHES.								7·10				
THIS WAS DURING VON RUNSTEDTS ARDENNES OFFENSIVE. KITE "GOT CHOP" JUST ON TOP OF US. HAD TO "SCRAM" TO AVOID HIM.								6·10				
					·40							
WANDERED OVER LEIPZIG BY MISTAKE. NUMEROUS FLAK HOLE IN A/C. - PRETTY NARROW SQUEAK								9·30				
					·30							
HOT TARGET. MORE FLAK DAMAGE. FW 190 SIGHTED AND COMBAT ENSUED. DAMAGE TO FIGHTER BY REAR-GUNNER								9·25				
38·45	26·55	4·00	-	141·20	157·45	4·55	38·00	162·15	6·00	68·05	63·30	39·4

YEAR: 1945 MONTH.	DATE.	AIRCRAFT. Type.	No.	PILOT, OR 1ST PILOT.	2ND PILOT, PUPIL, OR PASSENGER.	DUTY (INCLUDING RESULTS AND REMARKS).
—	—	—	—	—	—	— TOTALS BROUGHT FORWARD
FEBRUARY	1	LANCASTER	"A" 496	SELF	CREW	13TH OPERATION - SIEGEN - 12 x 1000 l
	2/3	"	"A" 496	SELF	CREW	14TH OPERATION - KARLSRUHE - CHOP NIGHT 1 x 4000 lbs + 12 CANS INCENDIARIES
	4	"	"A" 496	SELF	CREW	H.L.B. - WEATHER UNFIT.
	7/8	"	"A" 496	SELF	CREW	15TH OPERATION - DORTMUND EMMS CANAL - 12 x 1000 lbs.
	8/9	"	"A" 496	SELF	CREW	16TH OPERATION - PÖLITZ - CHOP AROUND 1 x 4000 lbs + 9x500 lbs.
	11	"	"M" 869	SELF	GR./CAPT HARRIS CREW	H. L. B.
	13/14	"	"A" 496	SELF	CREW	17TH OPERATION - DRESDEN - 1 x 2000 lb + 14 CANS INCENDIARIES
	14/15	"	"A" 496	SELF	CREW	18TH OPERATION - ROSITZ - 1 x 4000 lb + 9x500 lbs / DRESDEN STILL BURNING WELL
	19/20	"	"A" 496	SELF	CREW	19TH OPERATION - BOHLEN - SHAMBLES CONTROLLER CHOPPED. 1x4000 lb + 9 x 500 lbs
	21/22	"	"M" 869	SELF	CREW	20TH OPERATION - MITTELLAND CANAL (AGAIN) - 12 x 1000 lb 1/2 HR DELAY BRIGHT MOONLIGHT - CHOP ABOUT - WIZARD PRANG
	26	"	"Z" 372	F/LT WEBSTER	SELF, CREW	10/20 SORTIE CHECK
	27	"	"A" 496	SELF	CREW	FIGHTER AFFIL. & AIR-SEA FIRING

GRAND TOTAL [Cols. (1) to (10)]. 649 Hrs. 05 Mins.

TOTALS CARRIED FORWARD

	DAY.		NIGHT.		DAY.			NIGHT.			PASSENGER.	FLYING [Incl. in Cols. (1) to (10)].	
	DUAL.	PILOT.	DUAL.	PILOT.	DUAL.	1ST PILOT.	2ND PILOT.	DUAL.	1ST PILOT.	2ND PILOT.		DUAL.	PILOT.
	(1)	(2)	(3)	(4)	(5)	(6)	(7)	(8)	(9)	(10)	(11)	(12)	(13)
	38.45	26.55	4.00	-	141.20	157.45	4.55	38.00	162.15	6.00	68.05	63.30	39
FLEW IN MY DRESS UNIFORM, BUT NOT BECAUSE OF ANY SUPERSTITION									6.35				
2 COMBATS. 1 ME 110 & 1 JU. 88. HOLED BY CANON FIRE FROM ME. 110. TARGET AREA BRILLIANT AS DAY.									7.25				
						.55							
THE OLD MILK RUN AGAIN.									6.25				
VERY TOUGH TRIP. COMBAT WITH FW 190. CLAIMED 1 POSSIBLE. FLAK & FIGHTER DAMAGE TO AIR/C. KEN MANGOS KILLED ON THIS TRIP.									9.50				
						1.50							
MOST SUCCESSFUL RAID EVER CARRIED OUT. MOST DEVASTATING FIRES OF THE WAR. 35000 DIED THRO' SUFFOCATION ALONE. JERRY WHOLLY DISORGANISED.									-				
									9.50				
STICKY TARGET. TOO MUCH LOW CLOUD OVER TARGET. PASSED CHEMNITZ WHICH REST OF COMMAND WERE JUST PRANGING. FLAK DAMAGE. MID-UPPER GUNNER NEARLY KILLED. 1 RUNNING COMBAT WITH AN FW. 190. LOST HIM IN CLOUD.									-				
									9.30				
									-				
THICK CLOUD OVER TARGET. - DEAD LOSS. - MASTER BOMBER WAS KILLED TONIGHT.									8.15				
									-				
THE OLD MILK RUN AGAIN. BRIGHT MOON. FIGHTERS REAPED A HARVEST OF LANCASTERS TONIGHT. SAW ONE LANC. CRASH ON A JERRY AIRFIELD.									6.45				
						-			-				
						1.00							
						.50							
	38.45	26.55	4.00	.	141.20	162.20	4.55	38.00	226.50	6.00	68.05	63.30	57.

YEAR: 1945		AIRCRAFT.		PILOT, OR 1ST PILOT.	2ND PILOT, PUPIL, OR PASSENGER.	DUTY (INCLUDING RESULTS AND REMARKS).
MONTH.	DATE.	Type.	No.			
—	—	—	—	—	—	TOTALS BROUGHT FORWARD

SUMMARY FOR FEBRUARY 1945

		DAY	NIGHT
OPERATIONS	NIGHT		64·35
TRAINING	DAY	4·35	
	TOTALS.	4·35	64·35

TOTAL HOURS FOR FEB 69·10

H. F. Richardson F/L S/L O.C. "A" FLT

[signature] W/CDR OC. 44 SQN

GRAND TOTAL [Cols. (1) to (10)].
649 Hrs. 05 Mins.

TOTALS CARRIED FORWARD

DAY.		NIGHT.		DAY.			NIGHT.			PASSENGER.	FLYING [Incl. in Cols. (1) to (10)].	
DUAL.	PILOT.	DUAL.	PILOT.	DUAL.	1ST PILOT.	2ND PILOT.	DUAL.	1ST PILOT.	2ND PILOT.		DUAL.	PILOT.
(1)	(2)	(3)	(4)	(5)	(6)	(7)	(8)	(9)	(10)	(11)	(12)	(13)
38·45	26·55	4·00		141·20	162·20	4·55	38·00	226·50	6·00	68·05	63·30	57
38·45	26·55	4·00		141·20	162·20	4·55	38·00	226·50	6·00	68·05	63·30	57·,

YEAR: 1945 MONTH.	DATE.	AIRCRAFT. Type.	No.	PILOT, OR 1ST PILOT.	2ND PILOT, PUPIL, OR PASSENGER.	DUTY (INCLUDING RESULTS AND REMARKS).
—	—	—	—	—	—	TOTALS BROUGHT FORWARD
MARCH	2	LANCASTER	"H" 625	SELF	CREW	H.L.B.
	6/7	"	"P"	SELF	CREW	21st OPERATION – SASSNITZ 8 x 1000 lbs. – SHOT A SEARCHLIGHT OUT IN DENMARK ON RETURN FROM Δ
	20/21	"	"A" 496	SELF	CREW	22nd OPERATION – BOHLEN – 1x4000 + 9x500
	22	"	"F"	SELF	CREW	23rd OPERATION – HAMBURG D.R.A.G. 1x4000 + 11 x 500

SUMMARY FOR MARCH 1945.

		DAY	NIGHT
TRAINING	DAY	1·55	
OPERATIONS	NIGHT		23·25
			23·25

· 25·20 HRS TOTAL.

UNIT :- 44 SQDN.
DATE :- 31 · 3 · 45
SIGNATURE :- [signature] F/O

[signature] W/C O.C. 44 SQDN.

[signature] O.C. "A" FLIGHT

GRAND TOTAL [Cols. (1) to (10)].
674 Hrs. 25 Mins.

TOTALS CARRIED FORWARD

	DAY.		NIGHT.		DAY.			NIGHT.			...GER.	Cols. (1) to (10)].	
	DUAL.	PILOT.	DUAL.	PILOT.	DUAL.	1ST PILOT.	2ND PILOT.	DUAL.	1ST PILOT.	2ND PILOT.		DUAL.	PILOT
	(1)	(2)	(3)	(4)	(5)	(6)	(7)	(8)	(9)	(10)	(11)	(12)	(13)
	38·45	26·55	4·00	.	141·20	162·20	4·55	38·00	226·50	6·00	68·05	63·30	57
						1·55							
A SUCCESSFUL TRIP. ONLY ONE KITE LOST AND THAT WAS MY BEST FRIEND, "BUZZ" BOYLE, A CANADIAN FROM TORONTO. HE BLEW UP RIGHT IN FRONT OF ME. NO MORE HAPPY NIGHTS IN SKEGNESS WITH HIM.									9·20				
MUCH FLAK DAMAGE. MISSED A COLLISION OVER TARGET, BY HAIRSBREADTH.									8·50				
SHOT UP BY FLAK OVER TARGET AND ALSO OVER KIEL CANAL. FIRE IN FUSELAGE. EXTINGUISHED BY WIRELESS OPERATOR. STARBOARD OUTER PUT ON FIRE. O.K. AFTER GRAVINER SWITCH OPERATED. RETURNED ON 3 ENGINES. JERRY STILL HAS A LOT OF STING IN HIM.									5·15				
	38·45	26·55	4·00	.	141·20	164·15	4·55	38·00	250·15	6·00	68·05	63·30	61·

KEN WALKER'S WWII MEDALS

1 2 3 4 5

1. *1939-45 Star Bomber Command Medal*
2. *France-Germany Star*
3. *British Defence Medal*
4. *British 1939-45 Medal*
5. *NZ Defence Medal*

Made in the USA
Monee, IL
05 November 2026

67352885R10057